CalmDo Air Fryer Oven Cookbook for beginners

<u>Effortless Tasty Recipes for Your Calmdo Air Fryer Oven to Fry, Roast, Dehydrate, Bake and More</u>

By Johnson Amanda

© **Copyright 2020-All rights reserved.**

In no way is it legal to reproduce, duplicate, or transmit any part of this document by either electronic means or in printed format. Recording of this publication is strictly prohibited, and any storage of this material is not allowed unless with written permission from the publisher. All rights reserved.
The information provided herein is stated to be truthful and consistent, in that any liability, regarding inattention or otherwise, by any usage or abuse of any policies, processes, or directions contained within is the solitary and complete responsibility of the recipient reader. Under no circumstances will any legal liability or blame be held against the publisher for any reparation, damages, or monetary loss due to the information herein, either directly or indirectly.
Respective authors own all copyrights not held by the publisher.

Legal Notice:
This book is copyright protected. This is only for personal use. You cannot amend, distribute, sell, use, quote or paraphrase any part or the content within this book without the consent of the author or copyright owner. Legal action will be pursued if this is breached.

Disclaimer Notice:
Please note the information contained within this document is for educational and entertainment purposes only. Every attempt has been made to provide accurate, up to date and reliable, complete information. No warranties of any kind are expressed or implied. Readers acknowledge that the author is not engaging in the rendering of legal, financial, medical or professional advice.

By reading this document, the reader agrees that under no circumstances are we responsible for any losses, direct or indirect, which are incurred as a result of the use of information contained within this document, including, but not limited to, —errors, omissions, or inaccuracies.

TABLE OF CONTENTS

Book Description 1

Introduction for Calmdo Air Fryer Oven 3

    What Is It? 3

    How Does It Work? 4

    Various Functions 5

    Tips for Usage 7

    Matters Need Attention 7

    Cleaning & Maintenance 8

Chapter 1: Brunches 9 9

    Lemon Biscuits 9

    Brunch Hash 10

    Calmdo Air Fryer Oven Cookies 11

    Oat Sandwich Biscuits 12

    Chocolate Peanut Butter Cupcakes 13

    Cereal French Toast 15

    Brunch Frittata Recipe 16

    Avocado Egg 17

    Brunch Oats 18

Chapter 2: Beef, Pork, & Lamb 9 20

    Sugar Glaze Ham 20

    Corned Beef Rolls 21

    Lamb In Calm Do Air Fryer 22

    Salt And Black Pepper Steak 23

    Pork Chops 24

    Rump Steak 25

    Meat Patties... 26

    Teriyaki Glazed Steak... 27

    Country Style Ribs.. 29

Chapter 3: Fish & Seafood 9 30

    Ginger Garlic Salmon... 30

    Lemon Salmon.. 31

    Fish Taco.. 32

    Clams In The Air Fryer... 33

    Prawns Snack.. 35

    Coconut Cod Fillets... 36

    Fish Fillet In Air Fryer.. 37

    Prawns Snack.. 38

    Coconut Shrimp.. 39

Chapter 4: Chicken & Poultry 9 41

    Cornish Hen... 41

    Herb Roasted Hen... 42

    Chicken Milanese... 43

    Classic Blue Cheese Chicken Wings... 45

    Buffalo Wings.. 46

    Crispy Calm Do Air Fryer Chicken Leg.. 47

    Calm Do Air Fryer Chicken Breast... 49

    Chicken Broccoli.. 50

    Turkey In Calm Do Air Fryer... 51

Chapter 5: Vegan & Vegetarian 9 53

    Buffalo Cauliflower... 53

    Broiled Bananas..54

Stuffed Shells.. 55

Roasted Green Beans... 56

Spinach And Broccoli... 57

Blooming Onion... 58

French Fries... 60

Ginger Scones... 61

Potato Tots.. 62

Chapter 6: Soups, Stews and Broth 9..64

Chicken Gravy... 64

Steak And Mushroom Gravy.. 65

Mongolian Beef... 66

Lamb Chops In Yogurt... 67

Coconut Salmon Gravy... 68

Tomato Soup... 69

Vegetable Stew... 70

Madagascan Bean Stew.. 72

Mushroom Stew.. 73

Chapter 7: Beans and Eggs 9..75

Kidney Beans... 75

Garbanzo Beans.. 76

Navy Beans In Air Fryer... 77

Hard-Cooked Eggs.. 79

Eggs In A Hole... 80

Easy Breakfast Sandwich... 81

Air Fryer Crispy Chickpeas... 82

Omelet In Air Fryer.. 83

Mushroom Omelet In Air Fryer 84
Chapter 8: Desserts And Snacks 9 86
Zucchini Crisps 86

Apple Crisp 87

Walnut Brownies 88

Kale Chips 89

Simply Sweet Desert 90

Cheddar Biscuits 92

Banana And Chocolate Cups 93

Mixed Nuts 94

Bake Broccoli 95

Conclusion 97

Book Description

The CalmDo Air Fryer Oven is a remarkable appliance that makes it easy for all the people with on-to-go life to prepare some restaurant-style food easily in their kitchens and right to their dining tables.

It is a convection oven with 18 brilliant functions that include roast, dehydrate, reheat, bake, and much more.

The accessories that come with it are 10, and also it has a remarkable performance along with its great and attractive build in a style that occupies less space, and give some great result in a lesser amount of time.

If you are a busy person and find it hard to spend money to buy different types of equipment that are highly expensive to get the taste of crisp, bake, or roasted meals, then we highly recommend CalmDo Air Fryer Oven.

CalmDo Air Fryer Oven plays an important role in making healthy recipes for your whole family, without worrying too much.

Unlike, traditional frying and roasting it uses less oil and cook low calories meals that can be enjoyed by kids and adults.

In this book, we are covering the following:

- How Does It Work?
- Various Functions
- Tips For Usage
- Matters Need Attention

- Cleaning & Maintenance
- 72 Delicious Recipes
- Conclusion

The recipes are part of this guild to help the entire beginner prepare meals that are easy and scrumptious using a variety of functions available in the CalmDo Air Fryer Oven. Along with each recipe, snippets of nutritional information are given; so that the user keeps the calories intake right on track.

Now, lets the journey begin.

Introduction for Calmdo Air Fryer Oven

This part of the book discusses the basic protocol to understand the appliance along with its functions, work, and maintenance. Just like all the other air fryers the CalmDo Air Fryer Oven do air fryer uses an effective circulation technology that moves the hot air around the food to cook it to inside tenderness and outside crispness.

The final dish is an oil-free and delicious meal, which is not heavy on the health. The appliance is used to prepare oil-free meals every day with a hand-free cooking experience

What Is It?

The CalmDo Air Fryer Oven is more than just an air fryer it is a family-size cooking appliance that has enough space to prepare food for 3 -4 people .it's one-touch technology makes it easy to understand its vast functions and use it with ease. The 3 level racks make it easy to adjust different meals in one go.

It has 18 on touch functions that are listed below:

1. Chip
2. Cookie
3. Steak
4. Chicken
5. Shrimp
6. Fish
7. Bake
8. Rotisserie
9. Pizza
10. Dehydrate
11. Defrost
12. Reheat

13. Wing
14. Sausage
15. Cake
16. Nut
17. Vegetable
18. Lamb

You can roast, bake, dehydrate, and grill the food in the calmdo air fryer. Its stainless chamber with reflective structure make it attractive and makes the temperature uniform. It has detachable and transparent door.

The cleaning of equipment is effortless. The appliance come with 10 accessories as listed:

- Rotisserie Tong
- Rack
- Drip Tray
- Mesh Basket
- Rotisserie Rod,
- Skewer Rotisserie
- Rotisserie Cage
- Two Mesh Trays
- All The Accessories Are Dish Washer Safe

How Does It Work?

The CalmDo Air Fryer Oven uses a 360 degrees heated cyclonic air technology that circulate hot air all around the food to cook it to perfection. The in-built fan circulates the air around the food put inside the appliance; as a result very delicious, oil-free food is prepared.

Various Functions

- Brand   Calmdo
- Model Number: A-Air Fryer
- Product Dimensions: 36 x 32.5 x 32.5 cm; 8.56 Kilograms
- Capacity: 12 liters
- Item Weight: 8.56 kg

Power/Start–Stop Button

- The power button is use to turn ON and OFF the device.
- Once start the device the default temperature of device is 392°F and time is shown 15 minutes.
- Pressing the power button any time during cooking shut down the appliance.

Menu Button

Use to choose the initial function that user will use.

Rotation Button

It is use in the rotisserie mode. This can be used with any preset function.

Temperature Control Button

- It is use to control temperature between 41 degrees F to 149 degrees F
- Press this button and then press or to raise or low
- Dehydration range is given as 86°F to 176°F
- Defrost range is given as 86°F to 176°F
- Time Control Button
- as by its name its use to adjust the time
- The operating range is 1 minute to 24 hours

Other functions include:

- Raise time or temperature
- Lower time or temperature
- Function indicator

Tips for Usage

- The food or meal that is small in size needs lesser time.
- Turning or flipping the food is important to ensure equal cooking.
- It is recommended to oil spray the food to make it crisper.
- Place a calmdo oven-safe dish inside the air fryer when baking a cake.
- Do not place the calmdo air fryer oven on a wet area.
- Avoid using damaged plug and cords.
- Oversize food items should not be inserted in the CalmDo Air Fryer Oven.
- Do not use the appliance outdoor.

Matters Need Attention

Problem	Solution
Calmdo air fryer oven	Plug the power into an earthed wall socket.
Calmdo air fryer oven does not work	Adjust the cooking temperature
ingredients are not cooked	Redo
The ingredients are fried unevenly inside calmdo air fryer	Flip and turn the ingredients during cooking
Fried snacks are not crispy enough	Brush the food with oil
Calmdo air fryer oven give smoke	It is because of over greasing, so clean the accessories before using.

Cleaning & Maintenance

- First, unplug the CalmDo Air Fryer Oven from the electric socket.
- Make sure the appliance is cooled when started to clean.
- Use warm soapy water to clean all the accessories, rinse them under tap water.
- The accessories of the calmdo air fryer are dishwasher safe so it can be washed in the dishwasher.
- Clean the inside of the oven with a wet hot towel or absorbent sponge.
- Clean the air fryers outer body with a soft damp cloth.
- We do not recommend using detergent or cleaner.
- The door is removable and dishwasher safe as well.
- The oven is not intended to be used by children.
- It is necessary to check the voltage indicated.
- Do not submerge appliance
- in water
- Do not place the appliance on an unstable surface

Chapter 1: Brunches

Lemon Biscuits

Preparation Time: 12 Minutes
Cooking Time: 8 Minutes
Yield: 2 Servings

Ingredients
1/3 cup melted butter
½ cup caster sugar
2.5 cups self-rising flour
1 small lemon, zest, and juice
2 organic eggs
Oil spray, for greasing

Directions
Preheat the CalmDo Air Fryer Oven to 356 degrees F for 3 minutes.
In a bowl mix all the dry ingredients.
In a separate bowl whisk eggs and then add melted butter, lemon zest, and juice
Combine the ingredients of both bowls.
Knead it into soft nice dough.
Roll out the dough on a flat surface, and cut in the shape of biscuits.
Put the biscuits on the baking tray greased with oil spray.
Air fries it in CalmDo Air Fryer Oven for 8 minutes at 357 degrees F.
Once done, serve.

Nutrition Facts
Servings: 2
Amount per serving
Calories   1093
% Daily Value*
Total Fat 36.9g   47%

Saturated Fat 21.1g   105%
Cholesterol 245mg   82%
Sodium 283mg   12%
Total Carbohydrate 169.6g   62%
Dietary Fiber 4.2g   15%
Total Sugars 50.8g
Protein 22g

Brunch Hash

Preparation Time: 12 Minutes
Cooking Time: 20 Minutes
Yield: 4 servings

Ingredients
2 cups russet potatoes, peeled and cubed
3/4 cup kielbasa, precooked and cubed
1/2 cup corn
½ cup unsalted butter, melted
2 teaspoons paprika
Salt and black pepper, to taste
4 small yellow onion, peeled and chopped
½ cup carrots
1/2 cup green beans
Oil spray, for greasing

Directions
In a shallow bowl mix potatoes, kielbasa, onions, all the vegetables, and butter.
Then season it with pepper, salt, and paprika.
Select BAKE function of CalmDo Air Fryer Oven by adjusting the temperature to 400 degrees F and set time to 20 minutes.
Transfer the ingredients to a greased sheet pan and bake it for 20 minutes.

Shake the ingredient after 5 minutes of baking.
Once done, serve and enjoy.

Nutrition Facts
Servings: 4
Amount per serving
Calories   409
% Daily Value*
Total Fat 31.1g   40%
Saturated Fat 17.3g   87%
Cholesterol 90mg   30%
Sodium 688mg   30%
Total Carbohydrate 26.6g   10%
Dietary Fiber 5g 18%
Total Sugars 5.5g
Protein 8.9g

Calmdo Air Fryer Oven Cookies

Preparation Time: 20 Minutes
Cooking Time: 10 Minutes
Yield: 2 servings

Ingredients
1/3 cup of melted Butter
½ cup caster sugar
2 cups self-rising flour
½ teaspoon of vanilla essence
6 tablespoons of coconut milk
½ cup of cocoa powder
Oil spray, for greasing

Directions
Preheat the CalmDo Air Fryer Oven to 356 degrees F.

Take a bowl and mix cocoa, flour, and sugar in the bowl.
Pour in the melted butter and vanilla extract.
Then pour the coconut milk and mix thoroughly.
Roll out the mixture and use a cookie cutter to cut the cookies.
Place the cookies on to oil greased baking sheet.
Bake it into the CalmDo Air Fryer Oven for 10 minutes at 356 degrees F.
Serve and enjoy once cool.

Nutrition Facts

Servings: 2
Amount per serving
Calories   1070
% Daily Value*
Total Fat 45.7g   59%
Saturated Fat 30.8g   154%
Cholesterol 81mg   27%
Sodium 231mg   10%
Total Carbohydrate 159.8g   58%
Dietary Fiber 10.8g   39%
Total Sugars 52.4g
Protein 18.2g

Oat Sandwich Biscuits

Preparation Time: 28 Minutes
Cooking time: 15 minutes
Yield: 2 Servings

Ingredients
2 cups plain flour
1/3 cup butter
1/4 cup white sugar
1 egg, beaten
¼ cup desiccated coconut

1 cup oats
1 cup white chocolate
1 teaspoon of vanilla extract

Directions

Using a hand beater whisk the butter and sugar in a mixing bowl.
Crack eggs and whisk well
Then add coconut, oats, chocolate, and vanilla extract.
Now dump the all-purpose flour and mix all ingredients well.
Make biscuits shapes.
Roll the shapes in additional oats.
Place it in the CalmDo Air Fryer Oven baking sheet and bake for 15 minutes at 375 degrees F.
When the biscuits are cooked and cool, serve.

Nutrition Facts

Servings: 2
Amount per serving
Calories   1657
% Daily Value*
Total Fat 82.3g   106%
Saturated Fat 53.4g   267%
Cholesterol 181mg   60%
Sodium 341mg 15%
Total Carbohydrate 205.6g   75%
Dietary Fiber 12.3g   44%
Total Sugars 78.4g
Protein 28.3g

Chocolate Peanut Butter Cupcakes

Preparation Time: 12 Minutes
Cooking Time: 15 Minutes
Yield: 4 servings

Ingredients

2 cups chocolate cake mix
1 egg
2 egg yolks
1/3 cup olive oil
1/3 cup water, hot
1/3 cup sour cream
4 tablespoons of peanut butter
1 tablespoon of powdered sugar
Oil spray, for greasing

Directions

Take 20 muffin cups and double them up into 10
Take a non-stick pan and grease it with oil spray.
Then set it aside for further use.
In a shallow bowl combine egg, cake mix, egg yolks, olive oil, hot water, and sour cream.
Beat it until combined, using a hand beater.
In a separate bowl, mix peanut butter and white sugar.
Make about 10 small round balls with the hand of peanut butter.
Equally, pour 1/4 cup of the chocolate batter into 10 muffin cups.
Top muffin cups with prepared peanut butter balls.
Bake for 15 minutes at 300 degrees F in CalmDo Air Fryer Oven.
Then serve.

Nutrition Facts

Servings: 4
Amount per serving
Calories   694
% Daily Value*
Total Fat 45.6g   58%
Saturated Fat 10.5g   53%
Cholesterol 154mg   51%
Sodium 804mg 35%

Total Carbohydrate 68.3g 25%
Dietary Fiber 3g 11%
Total Sugars 36.1g
Protein 12.4g

Cereal French Toast

Preparation Time: 10 Minutes
Cooking Time: 20 Minutes
Yield: 2 servings

Ingredients
2 cups of coconut milk, sweetened
4 organic eggs
¼ teaspoon cinnamon
2 cups flake cereal, sugar-coated
6 slices brioche bread slices
Cooking spray, for greasing
Maple syrup, for serving

Directions
Whisk the egg in a bowl and pour in the milk and cinnamon.
Whisk egg in a bowl and add to coconut milk.
Crush the cereal and add it to a shallow bowl.
Dip the bread in the milk then put it into the cereal bowl.
Select the AIR Fry mode of CalmDo Air Fryer Oven and set the temperature to 425°F, for 20 minutes.
Press START.
Now oil sprays the baking Pan and places toast on it.
Bale it in batches according to the capacity of CalmDo Air Fryer Oven.
Once cooking is done, remove and let is serve with maple syrup.
Nutrition Facts
Servings: 2

Amount per serving
Calories   1737
% Daily Value*
Total Fat 97.1g   124%
Saturated Fat 53.6g   268%
Cholesterol 327mg   109%
Sodium 393mg   17%
Total Carbohydrate 203.9g   74%
Dietary Fiber 11.9g   43%
Total Sugars 107.5g
Protein 30.1g

Brunch Frittata Recipe

Preparation Time: 10 Minutes
Cooking Time: 15 Minutes
Yield: 2 servings

Ingredients
4-5 large organic eggs
2 Italian sausages, chopped
2 cherry tomatoes, chopped
Oil spray, for greasing
1/4 cup of parsley, chopped
1/2 cup of Parmesan cheese, per liking
Salt and Black Pepper, to taste

Directions
Preheat the CalmDo Air Fryer Oven at 365 degrees F.
Put the cherry tomatoes and sausage into a baking sheet or tray
Grease it with oil spray.
Bake it for 5 minutes.
Next, whisk the egg in a bowl

Add the parmesan cheese, salt, parsley, and pepper to the whisked eggs.
Take out the baking tray and transfer the ingredient to the egg bowl.
Transfer this bowl mixture to the cake pan and bake in the CalmDo Air Fryer Oven for 10 minutes at 350 degrees F.
Once the eggs get firm, serve and enjoy.

Nutrition Facts
Servings: 2
Amount per serving
Calories   696
% Daily Value*
Total Fat 54.9g   70%
Saturated Fat 21.9g   110%
Cholesterol 488mg   163%
Sodium 1366mg 59%
Total Carbohydrate 8.3g   3%
Dietary Fiber 1.7g   6%
Total Sugars 4.1g
Protein 43.5g

Avocado Egg

Preparation Time: 10 Minutes
Cooking Time: 15 Minutes
Yield: 2 servings

Ingredients
1 avocado, pitted
2 eggs organic
Salt and black pepper, to taste
4 bacon slices, cooked and chopped

Directions

Cut the avocado in half and remove the pit.
Scoop egg-sized flesh from the center.
Crack one egg in the center of the avocado sprinkle salt and black pepper on top.
Put the avocado onto the baking tray.
Put the tray in CalmDo Air Fryer Oven.
Set time to 10 minutes at 375 degrees F.
Press the start, and when the preheating is done, put the tray inside the oven.
Once the cooking cycle complete, take out and serve with a sprinkle of bacon bits.

Nutrition Facts
Servings: 2
Amount per serving
Calories   428
% Daily Value*
Total Fat 35.5g   46%
Saturated Fat 9.3g   47%
Cholesterol 42mg   14%
Sodium 917mg   40%
Total Carbohydrate 9.5g   3%
Dietary Fiber 6.7g   24%
Total Sugars 0.7g
Protein 19.6g

Brunch Oats

Preparation Time: 10 Minutes
Cooking Time: 15 Minutes
Yield: 2 servings

Ingredients
1 cup steel-cut oats

2 cups of coconut milk
1 apple, peeled and chopped
2 tablespoons of brown sugar
½ teaspoon of cinnamon

Directions
Take a shallow bowl and mix all the ingredients.
Pour it into a small heatproof bowl
Bake it in CalmDo Air Fryer Oven for 15 minutes at 360 degrees F.
Then serve as a delicious brunch

Nutrition Facts
Servings: 2
Amount per serving
Calories   726
% Daily Value*
Total Fat 58.9g   76%
Saturated Fat 50.7g   254%
Cholesterol 0mg   0%
Sodium 40mg   2%
Total Carbohydrate 51.5g 19%
Dietary Fiber 10.3g   37%
Total Sugars 28.4g
Protein 8.8g

Chapter 2: Beef, Pork, & Lamb 9

Sugar Glaze Ham

Preparation Time: 10 Minutes
Cooking Time: 20 Minutes
Yield: 6 Servings

Ingredients
3 pounds of ham, boneless
¼ cup pineapple juice
½ cup mustard
½ cup brown sugar
¼ teaspoon of cloves

Directions
In a medium bowl mix cloves, brown sugar, mustard, pineapple juice.
Mix it all well.
Insert the rotisserie shaft in the middle of the ham halves.
Then secure the forks.
Pour the prepared sauce over the ham.
Afterward, put it inside the CalmDo Air Fryer Oven.
Press Menu and choose the chicken function.
Once cooking is done, let it rest for a few minutes before serving.

Nutrition Facts
Servings: 6
Amount per serving
Calories   483
% Daily Value*
Total Fat 23.3g   30%
Saturated Fat 6.9g   34%
Cholesterol 129mg   43%

Sodium 2962mg 129%
Total Carbohydrate 26.6g 10%
Dietary Fiber 4.9g 18%
Total Sugars 13.7g
Protein 41g

Corned Beef Rolls

Preparation Time: 15 Minutes
Cooking Time: 21 Minutes
Yield: 4-5 Servings

Ingredients
10 egg rolls
1 pound corned beef, shredded
1.5 cup cabbage
Spicy mustard, as needed

Directions
Place the egg rolls on a flat surface, and place a small amount of corned beef in the center of the wrapper.
Now keep on adding a small amount of cabbage, and mustard.
Roll the wrappers and sealing the edges by brushing them with water.
Put the eggroll on a rack of CalmDo Air Fryer Oven.
Place it inside the CalmDo Air Fryer Oven.
Press menu to choose French fries and press start.
Let it cook for 7 minutes.
Cook it until brown.
Cook all the rolls in batches.
Once it's done, serve.

Nutrition Facts
Servings: 5

Amount per serving
Calories   374
% Daily Value*
Total Fat 15.8g   20%
Saturated Fat 6g   30%
Cholesterol 92mg   31%
Sodium 1187mg 52%
Total Carbohydrate 37.6g 14%
Dietary Fiber 3.1g   11%
Total Sugars 3.7g
Protein 19.1g

Lamb In Calm Do Air Fryer

Preparation Time: 15 Minutes
Cooking Time: 20 Minutes
Yield: 6 Servings

Ingredients
2 limes, divided
1 tablespoon of lemon zest
Salt and black pepper, to taste
6 garlic cloves, minced
¼ cup mint, fresh
½ cup olive oil
12 lamb chops

Directions
Take a bowl and mix lemon zest, salt, pepper, garlic cloves, mint, and olive oil
Rub this over the lamb and marinate it for a few hours.
Place the lamb on the CalmDo Air Fryer Oven rack.
Press menu and select steak function.
Cook it for 20 minutes by pressing start.

Flip the lamb halfway through.
Once the lamb is cooked, serve, and enjoy.

Nutrition Facts
Servings: 6
Amount per serving
Calories   1367
% Daily Value*
Total Fat 64.8g   83%
Saturated Fat 19.5g   98%
Cholesterol 589mg   196%
Sodium 499mg   22%
Total Carbohydrate 1.5g   1%
Dietary Fiber 0.4g   1%
Total Sugars 0.1g
Protein

Salt And Black Pepper Steak

Preparation Time: 20 Minutes
Cooking Time: 12 Minutes
Yield: 3 Servings

Ingredients
3 sirloin steaks
Salt and black pepper, to taste
6 tablespoons of melted butter

Directions
Rub the steak with salt and black pepper.
Brush the steak with melted butter from both sides.
Oil sprays the CalmDo Air Fryer Oven basket.
Put the steak inside a basket.
Set the timer to 12 minutes at 392 degrees F.

Once done, serve and enjoy.
remember to cook it in batches.

Nutrition Facts
Servings: 3
Amount per serving
Calories   625
% Daily Value*
Total Fat 37.2g   48%
Saturated Fat 19.9g   100%
Cholesterol 264mg   88%
Sodium 313mg   14%
Total Carbohydrate 0g   0%
Dietary Fiber 0g 0%
Total Sugars 0g
Protein 69g

Pork Chops

Preparation Time: 20 Minutes
Cooking Time: 20 Minutes
Yield: 2 Servings

Ingredients
50 grams of brown sugar
Salt and black pepper, to taste
1 bay leaf
2 tablespoons of bourbon
4 pork chops

Directions
Take a saucepan and heat it over medium flame.
Add brown sugar, salt, pepper, bay leaf, and bourbon in a saucepan.
Heat it over a medium flame.

Take a baking dish and place pork and brine in it.
Coat the pork chops with brine well.
Then set the timer at 400 degrees F for 20 minutes.
Put the dish inside the CalmDo Air Fryer Oven.
Remember to flip the pork chops halfway through
Once done, serve

Nutrition Facts
Servings: 2
Amount per serving
Calories   639
% Daily Value*
Total Fat 39.8g   51%
Saturated Fat 14.9g   75%
Cholesterol 138mg   46%
Sodium 119mg 5%
Total Carbohydrate 24.6g 9%
Dietary Fiber 0g 0%
Total Sugars 24.3g
Protein 36g

Rump Steak

Preparation Time: 20 Minutes
Cooking Time: 10 Minutes
Yield: 2 Servings

Ingredients
1 pound of rump steak
1 onion, sliced
1 green bell pepper, sliced
Salt and black pepper, to taste
1 cup parmesan cheese
Hoagies roll, as needed

Directions

Take CalmDo Air Fryer Oven mesh tray and place steak, bell pepper, and onion.
Season it with salt and black pepper.
Put it inside CalmDo Air Fryer Oven.
Let it Bake for 10 minutes at 338 degrees F.
Once done, place it on a hoagie roll.
Sprinkle shredded parmesan on top.
Serve.

Nutrition Facts

Servings: 2
Amount per serving
Calories   827
% Daily Value*
Total Fat 33.3g   43%
Saturated Fat 12.5g   63%
Cholesterol 60mg   20%
Sodium 1014mg 44%
Total Carbohydrate 34.7g 13%
Dietary Fiber 5.5g   20%
Total Sugars 7.3g
Protein 103.5g

Meat Patties

Preparation Time: 15 Minutes
Cooking Time: 15 Minutes
Yield: 4 Servings

Ingredients

2 pounds of grounded beef
Black pepper, to taste
Salt, to taste

2 tablespoons of olive oil
½ cup shallots
2 green peppers, chopped
¼ teaspoon of cumin
Oil spray, for greasing

Directions

Take a bowl and mix olive oil, ground beef, salt, cumin, black pepper, shallots, and green pepper.
Prepare the meat patties with wet hands
Oil sprays the beef patties and then layers them on to mesh tray of CalmDo Air Fryer Oven.
Bake at 375 degrees for 15 minutes in CalmDo Air Fryer Oven, until golden brown for the top.

Nutrition Facts

Servings: 4
Amount per serving
Calories   187
% Daily Value*
Total Fat 13.3g   17%
Saturated Fat 3.6g   18%
Cholesterol 37mg   12%
Sodium 77mg   3%
Total Carbohydrate 6.2g   2%
Dietary Fiber 1g 4%
Total Sugars 1.4g
Protein 11.5g

Teriyaki Glazed Steak

Preparation Time: 15 Minutes
Cooking Time: 25 Minutes
Yield: 2 Servings

Ingredients
2 beef Steak

Teriyaki Glaze Ingredients
1/4 cup Soy Sauce
½ cup Japanese cooking wine
1/4 cup Brown Sugar
2 tablespoons Lime Juice
1/3 cup Orange Juice
1 teaspoon Ginger, ground
1 teaspoon of minced garlic

Directions
Mix all the glaze ingredients in a saucepan and let it cook for 5-8 minutes.
Set aside for cooling.
Now coat the steak with glaze and let it sit for 2 hours
Now place it on a baking tray and place it inside CalmDo Air Fryer Oven.
Set the timer of v CalmDo Air Fryer Oven at 392 degrees F, and cook for 15 minutes.
Once done, serve
Enjoy.

Nutrition Facts
Servings: 2
Amount per serving
Calories   589
% Daily Value*
Total Fat 21.8g   28%
Saturated Fat 8.1g   41%
Cholesterol 128mg   43%
Sodium 1975mg 86%
Total Carbohydrate 25.6g 9%
Dietary Fiber 0.5g   2%
Total Sugars 21.6g
Protein 68.7g

Country Style Ribs

Preparation Time: 12 Minutes
Cooking Time: 12 Minutes
Yield: 4 Servings

Ingredients

8 country-style pork ribs, trimmed excess fat
4 tablespoons cornstarch
4 tablespoons coconut oil
4 teaspoon dry mustard
½ teaspoon thyme
½ teaspoon garlic powder
1 teaspoon dried marjoram
Salt and black pepper, to taste

Directions

Take a bowl and combine pork with cornstarch, coconut oil, dry mustard, thyme, garlic powder, marjoram, salt, and black pepper. Rub the pork well and let it marinate for a few hours.
Now put the ribs in a CalmDo Air Fryer Oven basket and roast it for 12 minutes at 392 degrees F.
Once done, serve.

Nutrition Facts

Servings: 4
Amount per serving
Calories   1118
% Daily Value*
Total Fat 38g   49%
Saturated Fat 19.8g   99%
Cholesterol 486mg   162%
Sodium 381mg   17%
Total Carbohydrate 8.9g   3%
Dietary Fiber 0.7g   3%
Total Sugars 0.3g
Protein 175.2g

Chapter 3: Fish & Seafood 9

Ginger Garlic Salmon

Preparation Time: 10 Minutes
Cooking Time: 12 Minutes
Yield: 2 Servings

Ingredients
4 salmon fillets
Salt and black pepper, to taste
4 tablespoons of melted butter
4 cloves of garlic, minced
1-inch ginger, minced
Red chili flakes
1 tablespoon of coconut amino

Directions
Take a mesh basket and cover it with Aluminum foil.
Put Salmon in the mesh basket.
In a bowl combine garlic, ginger, coconut amino, salt black, red chili flakes, and pepper butter.
Pour the bowl mixture over the salmon.
Turn on the CalmDo Air Fryer Oven and set the temperature to 10 minutes at 350 degrees Fahrenheit.
Put the mesh basket into the CalmDo Air Fryer Oven.
Let it bake for 10 minutes.
Turn on broil for 2 minutes at 400 degrees Fahrenheit to bake the top of the Salmon.
Afterward, take out and serve.
Enjoy.

Nutrition Facts
Servings: 2

Amount per serving
Calories 684
% Daily Value*
Total Fat 45.1g 58%
Saturated Fat 17.7g 89%
Cholesterol 218mg 73%
Sodium 322mg 14%
Total Carbohydrate 2g 1%
Dietary Fiber 0.1g 1%
Total Sugars 0.1g
Protein 69.7g

Lemon Salmon

Preparation Time: 10 Minutes
Cooking Time: 10 Minutes
Yield: 3 servings

Ingredients
1 pound salmon fillet
¼ teaspoon lemons, zest, and juice
2 tablespoons olive oil
¼ teaspoon of turmeric
1/3 teaspoon of cumin
½ teaspoon of red chili flakes
1/3 teaspoon of oregano
Salt and black pepper

Directions
Take a mesh basket and cover it with Aluminum foil.
Rub the salmon fillet with the listed ingredients.
Put Salmon in the mesh basket.
Turn on the CalmDo Air Fryer Oven and set the temperature to 10 minutes, at 350 degrees Fahrenheit.

Put the mesh basket into the CalmDo Air Fryer Oven let it bake for 8 minutes.

Then turn on broil for 2 minutes at 400 degrees Fahrenheit to bake the top of the Salmon.

Afterward, take out and serve.

Enjoy.

Nutrition Facts

Servings: 3
Amount per serving
Calories   282
% Daily Value*
Total Fat 18.8g   24%
Saturated Fat 2.7g   13%
Cholesterol 67mg   22%
Sodium 67mg   3%
Total Carbohydrate 0.4g   0%
Dietary Fiber 0.2g   1%
Total Sugars 0g
Protein 29.4g

Fish Taco

Preparation Time: 10 Minutes
Cooking Time: 10 Minutes
Yield: 2 servings

Ingredients

12ounces cod fish fillet
1 cup Panko crumbs
1 cup coleslaw

Tempura batter ingredients

1 cup flour
1 1/3 cup corn starch
1 cup of water

Directions

The first step is to cut the fish fillets into 3-ounce pieces.
Take a bowl and combine all the tempura batter ingredients.
Dip the fish fillet into the tempura batter and then coat it with Panko bread crumbs.
Place the fish fillet in a CalmDo Air Fryer Oven baking tray and place it in the oven.
Press the menu and choose French fries.
Adjust the cooking time to 10 minutes.
Once the cooking time complete take out the fish and serve with coleslaw.

Nutrition Facts

Servings: 2
Amount per serving
Calories   5360
% Daily Value*
Total Fat 283.9g 364%
Saturated Fat 49.2g   246%
Cholesterol 447mg   149%
Sodium 5811mg 253%
Total Carbohydrate 614.2g   223%
Dietary Fiber 13.7g   49%
Total Sugars 0.2g
Protein 127.4g

Clams In The Air Fryer

Preparation Time: 10 Minutes
Cooking Time: 3 Minutes
Yield: 6 servings

Ingredients

1 cup bread crumb

½ cup parmesan
1.2 cup parsley
½ teaspoon of lemon zest
6 tablespoons of butter
2 garlic cloves
1 dozen clams
1 lemon, wedges

Direction
Take a Shallow bowl and combine Parmesan cheese, breadcrumbs, lemon zest garlic, parsley, and melted butter.
Mix the ingredients well and then place this mixture on top of the exposed clams.
Season it with salt and black pepper.
Press start and select the French fries function of the CalmDo Air Fryer Oven.
Adjust the cooking time to 2-3 minutes.
cook it in batches ,
Once done, serve it with lemon wedges

Nutrition Facts
Servings: 6
Amount per serving
Calories   749
% Daily Value*
Total Fat 20.5g   26%
Saturated Fat 8.6g   43%
Cholesterol 287mg   96%
Sodium 668mg   29%
Total Carbohydrate 33.7g 12%
Dietary Fiber 1.2g   4%
Total Sugars 1.3g
Protein 100.8g

Prawns Snack

Preparation Time: 10 Minutes
Cooking Time: 6 Minutes
Yield: 2 servings

Ingredients

10 fresh king prawns
1 tablespoon of wine vinegar
4 tablespoons of mayonnaise
1 teaspoon of ketchup
1 teaspoon chili flakes
1/3 tsp of sea salt
1/3 tsp of ground black pepper
1 teaspoon of chili powder

Directions

Preheat the CalmDo Air Fryer Oven at 360 degrees F for 3minutes. Mix all the ingredients in a shallow bowl and coat the prawns well. Transfer it to a baking tray and bake it in CalmDo Air Fryer Oven for 6 minutes.
Once done, serve.

Nutrition Facts

Servings: 2
Amount per serving
Calories   612
% Daily Value*
Total Fat 28.8g   37%
Saturated Fat 1.5g   7%
Cholesterol 8mg   3%
Sodium 662mg   29%
Total Carbohydrate 21.9g 8%
Dietary Fiber 1.1g   4%
Total Sugars 3.1g

Protein 71.8g

Coconut Cod Fillets

Preparation Time: 10 Minutes
Cooking Time: 12 Minutes
Yield: 2 servings

Ingredients
Salt and black pepper, to taste
1 pound of cod fish fillet
1 cup of coconut milk
¼ teaspoon of paprika, smoked

Directions
First, preheat the CalmDo Air Fryer Oven to 375 degrees F.
In a bowl and combine salt, black pepper, coconut milk, paprika, and marinate fish for a few hours.
Now put the fish in a CalmDo Air Fryer Oven baking sheet and cook for about 12 minutes, at 375 degrees.
 Serve and enjoy.
Enjoy.

Nutrition Facts
Servings: 2
Amount per serving
Calories   515
% Daily Value*
Total Fat 30.6g   39%
Saturated Fat 25.8g   129%
Cholesterol 125mg   42%
Sodium 195mg 8%
Total Carbohydrate 6.8g   2%
Dietary Fiber 2.8g   10%

Total Sugars 4g
Protein 54.6g

Fish Fillet In Air Fryer

Preparation Time: 10 Minutes
Cooking Time: 12 Minutes
Yield: 2 Servings

Ingredients
2 cups breadcrumbs
Salt, to taste
Black pepper, to taste
4 tablespoons fresh parsley
175g firm white fish fillet
1 cup plain all-purpose flour
2 eggs, whisked

Ingredients for Sauce
½ cup mayonnaise
1 tablespoon of capers, drained
2 jalapeños, chopped
1 tablespoon of lemon juice
Pinch of chili flakes
Salt, to taste

Directions
Preheat the CalmDo Air Fryer Oven to 375 degrees F for 3 minutes.
Combine the salt, pepper, breadcrumbs, and parsley in the bowl.
Whisk the egg in a separate bowl
Put the flour on a flat plate.
Dip the fillets first in the flour, then the egg, and then in crumb mixture.

Once all the fillets are coated, put it on a baking sheet and place it inside CalmDo Air Fryer Oven, for about 12 minutes.
Meanwhile, combine all sauce ingredients in a bowl
Remember to flip the fish halfway through.
Once cooking is complete, serve with sauce.

Prawns Snack

Preparation Time: 10 Minutes
Cooking Time: 18 Minutes
Yield: 4 servings

Ingredients
12 king prawns, fresh
1 tablespoon of wine vinegar
4 tablespoons of mayonnaise
1 teaspoon of ketchup
1 teaspoon chili flakes
Salt and black pepper, to taste
½ teaspoon of chili powder

Directions
Preheat your CalmDo Air Fryer Oven to 320 degrees F for few minutes.
In a medium bowl, mix all the spices and add prawns.
Place the prawns into the CalmDo Air Fryer Oven frying basket, and cook for 6 minutes.
Then serve.

Nutrition Facts
Servings: 4
Amount per serving
Calories   139
% Daily Value*

Total Fat 6.1g 8%
Saturated Fat 1.1g 5%
Cholesterol 143mg 48%
Sodium 333mg 14%
Total Carbohydrate 9.8g 4%
Dietary Fiber 0.4g 1%
Total Sugars 1.5g
Protein 15.2g
Vitamin D 0mcg

Coconut Shrimp

Preparation Time: 10 Minutes
Cooking Time: 8 Minutes
Yield: 2 servings

Ingredients
1 cup sour cream
2 cups pineapple chunks, (liquid reserved)
2 egg whites
1 cup cornstarch
2/3 cup sweetened coconut
1 cup breadcrumbs
1-1/2 pound large shrimp, thawed
Olive oil, for misting

Directions
Take a shallow bowl and combine the pineapple and sour cream in it.
In a small bowl beat the egg whites and add reserved pineapple liquid.
Place the cornstarch on a flat tray.
Mix the coconut and breadcrumbs in a separate bowl.
Dip the shrimp first into cornstarch, then into the egg wash

Then dredge it into a crumb mixture.
Place the shrimp in the CalmDo Air Fryer Oven basket according to space.
 Spray it with oil.
Let it cook for 8 minutes at 375 degrees F in CalmDo Air Fryer Oven.
Serve with prepared dipping pineapple sauce.

Nutrition Facts

Servings: 2
Amount per serving
Calories   1139
% Daily Value*
Total Fat 43.2g   55%
Saturated Fat 24.6g   123%
Cholesterol 375mg   125%
Sodium 786mg 34%
Total Carbohydrate 132.2g   48%
Dietary Fiber 7.7g   28%
Total Sugars 21.7g
Protein 58.9g

Chapter 4: Chicken & Poultry 9

Cornish Hen

Preparation Time: 10 Minutes
Cooking Time: 40-80 Minutes (In Batch)
Yield: 2 Servings

Ingredients
Salt and ground black pepper, to taste
2 Cornish hens
2 teaspoon of garlic powder
2 rosemary leaves, fresh and chopped

Directions
Take a bowl and mix salt, black pepper, garlic powder, and rosemary leaves.
Rub the hens with the prepared rub.
Place the hens on the rotisserie shaft and secure the forks.
Place the hens inside Calm Do Air Fryer Oven.
Remember to tuck and tie the wings to keep them in place.
Press the menu and choose the chicken function.
Set the cooking to 40 minutes, and press start.
You can cook the chicken in a batch.
Once done, serve.

Nutrition Facts
Servings: 2
Amount per serving
Calories   157
% Daily Value*
Total Fat 4.3g   5%
Saturated Fat 1.1g   5%
Cholesterol 117mg   39%

Sodium 70mg   3%
Total Carbohydrate 2g   1%
Dietary Fiber 0.3g   1%
Total Sugars 0.7g
Protein 26.1g

Herb Roasted Hen

Preparation Time: 10 Minutes
Cooking Time: 40 Minutes
Yield: 2 Servings
1 Hour to Marinate

Ingredients
1 Cornish hen
4 spring Thyme
2 spring of Sage
1 spring of Rosemary
Salt and black pepper, to taste
2 cloves of garlic, minced
1-inch of ginger, minced
4 tablespoons of olive oil

Directions
Take a blender and pulse together salt, garlic, ginger, black pepper, olive oil, sage, rosemary, and thyme.
Rub this mixture over the hen.
Let the hen rest in the refrigerator for 1 hour to marinate.
Put the hen on the rotisserie shaft and secure the forks.
Tuck and tie the wings to keep them in place.
Place the hen inside Calm Do Air Fryer.
Press the menu and choose the chicken function.
Set the cooking time to 40 minutes.
Once it's done, serve and enjoy.

Nutrition Facts
Servings: 2
Amount per serving
Calories   318
% Daily Value*
Total Fat 30.2g   39%
Saturated Fat 4.5g   23%
Cholesterol 58mg   19%
Sodium 35mg   2%
Total Carbohydrate 1g   0%
Dietary Fiber 0.1g   0%
Total Sugars 0g
Protein 13g

Chicken Milanese

Preparation Time: 15 Minutes
Cooking Time: 15 Minutes
Yield: 2 Servings

Ingredients
2 cups Panko bread crumbs
1 tablespoon of garlic powder
Salt and black pepper, to taste
½ cup of Parmesan Cheese
5 chicken cutlets
2 eggs, whisked

Ingredients for Salad
2 cups arugula
Salt and black pepper to taste
2 tablespoons of lemon juice
1 tablespoon of vinegar
4 tablespoons of olive oil

Directions

Whisk eggs in a bowl and set aside for further use.
Take a large bowl and combine garlic, salt, pepper, and breadcrumbs.
Rub the chicken cutlets with salt and black pepper and dip the cutlets into the egg wash.
Then put the chicken into the bread crumb mixture to coat it evenly.
Place the chicken onto the oven Rack and put it inside a Calm Do Air Fryer Oven.
Press the manual and chose the bake function.
Adjust the time to 15 minutes and then press start.
Take a bowl and combine olive oil, salt, pepper, lemon juice, vinegar.
Put the arugula into the dressing bowl and coat it well.
Top the chicken with the salad and serve with shaved Parmesan cheese.

Nutrition Facts

Servings: 2
Amount per serving
Calories   1582
% Daily Value*
Total Fat 74.4g   95%
Saturated Fat 20.3g   101%
Cholesterol 519mg   173%
Sodium 1566mg 68%
Total Carbohydrate 83.8g   30%
Dietary Fiber 5.7g   20%
Total Sugars 8.8g
Protein 140.4g

Classic Blue Cheese Chicken Wings

Preparation Time: 15 Minutes
Cooking Time: 20 Minutes
Yield: 3 Servings
Refrigerate for 1 hour

Ingredients

2 garlic cloves minced
2 teaspoon of mustard
One teaspoon of Paprika powder
Salt and black pepper to taste
2 tablespoons of olive oil
Oil spray, for greasing
12 chicken wings
1 cup blue cheese, for coating

Directions

Take a bowl and combine minced garlic, salt, pepper, mustard, paprika, and olive oil.
Coat the chicken wings with the marinade and put it in the refrigerator for 1 hour.
Take mesh basket and grease with oil spray.
Put the Calm Do Air Fryer Oven chicken wings into the mesh basket.
Set it to 20 minutes at 356 degrees Fahrenheit.
Put the mesh basket inside the Calm Do Air Fryer Oven.
Cook it for 20 minutes, until crispy and golden brown.
Do it in batches.
Serve with a rub of blue cheese and enjoy.

Nutrition Facts

Servings: 3
Amount per serving
Calories   1361
% Daily Value*
Total Fat 66.4g   85%

Saturated Fat 21.7g 108%
Cholesterol 554mg 185%
Sodium 1130mg 49%
Total Carbohydrate 1.8g 1%
Dietary Fiber 0.3g 1%
Total Sugars 0.4g
Protein 179.1g

Buffalo Wings

Preparation Time: 15 Minutes
Cooking Time: 35 Minutes
Yield: 4 Servings

Ingredients
2 pounds of chicken wings
1 teaspoon of sea salt
1 teaspoon garlic powder
1/2 teaspoon ground black pepper
1/2 teaspoon baking powder
Olive oil, spray

For the Buffalo Wing Sauce
1/2 cup hot sauce, like Frank's
2 tablespoons butter

Directions
Preheat the CalmDo Air Fryer Oven to 390 degrees F.
Take a bowl and combine sea salt, black pepper, garlic powder, and baking powder.
Rub this mixture over the chicken wings.
Grease it with the oil spray.
Now put the wings in a CalmDo Air Fryer Oven baking tray.
Set timer to 30 minutes.

Meanwhile prepare buffalo sauce and for that cook all the sauce ingredients in a cooking pan until all the ingredients melted.
Once the chicken is cooked, take it out and let it rest for 5 minutes. Coat the cooked chicken with the sauce and serve.
Enjoy.

Nutrition Facts
Servings: 4
Amount per serving
Calories   489
% Daily Value*
Total Fat 22.7g   29%
Saturated Fat 8.3g   41%
Cholesterol 217mg   72%
Sodium 1466mg 64%
Total Carbohydrate 1.5g   1%
Dietary Fiber 0.2g   1%
Total Sugars 0.5g
Protein 66g
Vitamin D 4mcg

Crispy Calm Do Air Fryer Chicken Leg

Preparation Time: 15 Minutes
Cooking Time: 30 Minutes
Yield: 4 Servings

Ingredients
8 chicken leg pieces
3 cups of buttermilk
2 tablespoons of hot sauce
2 eggs, whisked
2 cups all-purpose flour
½ cup of corn starch

Salt and black pepper, to taste
2 teaspoons of onion powder
2 teaspoons of paprika
2 teaspoons of garlic powder
Oil spray, for spraying

Directions

Take a large metal bowl and mix eggs with buttermilk, salt, black pepper, 1 teaspoon paprika, 1 teaspoon of garlic powder, and hot sauce.
Put leg pieces in the marinade and refrigerate for 6 hours.
Now start to prepare chicken
Take a bowl and combine flour, corn starch, salt, 1 teaspoon of paprika, onion powder, and 1 teaspoon of garlic powder.
Dip the chicken in egg wash.
Dredge the chicken into flour.
Layer the chicken on to oil grease rack.
Spray the chicken with oil spray.
 Place it on a calm do baking tray.
Cook it in a CalmDo Air Fryer Oven at 350 degrees F for 30 minutes. After 15 minutes, flip the chicken pieces and oil grease with cooking spray. Complete the cooking cycle and then serve.

Nutrition Facts

Servings: 4
Amount per serving
Calories   971
% Daily Value*
Total Fat 26.4g   34%
Saturated Fat 7.8g   39%
Cholesterol 349mg   116%
Sodium 668mg 29%
Total Carbohydrate 77.4g 28%
Dietary Fiber 2.3g   8%
Total Sugars 10.1g
Protein 100.3g

Calm Do Air Fryer Chicken Breast

Preparation Time: 15 Minutes
Cooking Time: 15 Minutes
Yield: 4 Servings

Ingredients

1 large egg, whisked
1/4 cup all-purpose flour
Salt and black pepper, to taste
3/4 cup Panko bread crumbs
1/3 cup freshly grated Parmesan
2 tablespoons lemon zest
1 tablespoon dried oregano
1/2 teaspoon cayenne pepper
2 chicken breasts, boneless and skinless

Directions

Take a shallow bowl and whisk the egg in it.
Take another bowl and add flour, lemon zest, oregano, and cayenne pepper.
Then add salt and black pepper to flour and egg.
Dip the chicken into the egg and then into the flour mixture.
Put the chicken in Panko bread crumbs.
Coat the chicken well.
Put it in a CalmDo Air Fryer Oven basket and cook it at 400 degrees F for 10-15 minutes.
Flip the chicken halfway through.
Cook until golden from the top.
Serve by sprinkling parmesan on top.

Nutrition Facts

Servings: 4
Amount per serving
Calories   279
% Daily Value*
Total Fat 8.5g   11%

Saturated Fat 2.5g 13%
Cholesterol 113mg 38%
Sodium 251mg 11%
Total Carbohydrate 22.2g 8%
Dietary Fiber 1.9g 7%
Total Sugars 1.6g
Protein 27.2g

Chicken Broccoli

Preparation Time: 10 Minutes
Cooking Time: 20 Minutes
Yield: 2 Servings

Ingredients

1 pound chicken breast, cut into pieces
¼ pound broccoli, florets
1 medium onion, sliced thick
4 tablespoons olive oil or grape seed oil
1/4 teaspoon garlic powder
½ tablespoon ginger, minced
½ tablespoon soy sauce
2 teaspoons sesame seed oil
2 teaspoons rice vinegar
4 teaspoons hot sauce
Salt and black pepper, to taste

Directions

Take a medium or large bowl and add chicken breast, onions, and broccoli.
Mix the ingredients and set them aside for further use.
In a small bowl, mix oil, salt, black pepper, ginger, soy sauce, garlic, rice vinegar, and sesame oil.
Mix it all well.
Add it to broccoli and chicken mixture.

Put the chicken in a basket covered with foil, and place it in CalmDo Air Fryer Oven.
Turn on the air fryer at 360 degrees for 20 minutes.
Shake the basket a few times during cooking.
Once done, serve with hot sauce.

Nutrition Facts
Servings: 2
Amount per serving
Calories   592
% Daily Value*
Total Fat 38.6g   49%
Saturated Fat 4.7g   23%
Cholesterol 145mg   48%
Sodium 616mg 27%
Total Carbohydrate 10.6g 4%
Dietary Fiber 2.9g   10%
Total Sugars 3.6g
Protein 50.8g

Turkey In Calm Do Air Fryer

Preparation Time: 10 Minutes
Cooking Time: 30 Minutes
Yield: 4 Servings

Ingredients
2 pounds turkey breast, rib removed
1 tablespoon olive oil
½ teaspoon salt
14 tablespoons dry turkey seasoning

Directions
Rub the turkey with olive oil and season the sides with salt and turkey seasoning.

Preheat the CalmDo Air Fryer Oven at 350 degrees F.
Transfer the turkey to a baking tray and put it in a CalmDo Air Fryer Oven and cook for 30 minutes by selecting chicken.
Let it rest for 10 minutes, before serving.

Nutrition Facts

Servings: 4
Amount per serving
Calories   266
% Daily Value*
Total Fat 7.3g   9%
Saturated Fat 1.3g   6%
Cholesterol 98mg   33%
Sodium 2593mg 113%
Total Carbohydrate 9.6g   3%
Dietary Fiber 1.1g   4%
Total Sugars 8g
Protein 38.7g

Chapter 5: Vegan & Vegetarian 9

Buffalo Cauliflower

Preparation Time: 14 Minutes
Cooking Time: 20 Minutes
Yield: 4 Servings

Ingredients
1 head of cauliflower
Salt and black pepper, to taste
1 tablespoon of olive oil
1 tablespoon of lemon juice
1 cup buffalo sauce

Directions
Rinse and pat dry the cauliflower
Cut the cauliflower in florets
In a bowl combine buffalo sauce, salt, lemon juice, and pepper along with oil
Mix well and dredge cauliflower in the mixture
Transfer the cauliflower florets in a mesh basket and place them in CalmDo Air Fryer Oven.
Set timer to 20 minutes at 310 degrees F
Once done, serve.

Nutrition Facts
Servings: 4
Amount per serving
Calories   48
% Daily Value*
Total Fat 3.6g   5%
Saturated Fat 0.5g   3%
Cholesterol 0mg   0%

Sodium 21mg 1%
Total Carbohydrate 3.6g 1%
Dietary Fiber 1.7g 6%
Total Sugars 1.7g
Protein 1.3g

Broiled Bananas

Preparation Time: 15 Minutes
Cooking Time: 10 Minutes
Yield: 4 Servings

Ingredients
2 tablespoons sugar, dark brown
1 tablespoon cinnamon, ground
4 bananas firm and cut lengthwise

TOPPINGS Ingredients
½ cup Walnuts, chopped
½ cup Whipped cream
Chocolate syrup, as needed

Directions
Take a bowl and mix sugar, cinnamon, and coat the bananas with it
Put the banana on a baking sheet that is covered with a sheet pan. Place it in CalmDo Air Fryer Oven
Set timer to 10 minutes at 310 degrees F
Once the top of the banana is caramelized take it out, and allow it to get cool
Then serve with the toppings.

Nutrition Facts
Servings: 4
Amount per serving

Calories 299
% Daily Value*
Total Fat 14g   18%
Saturated Fat 3.5g   17%
Cholesterol 17mg   6%
Sodium 14mg   1%
Total Carbohydrate 42.7g 16%
Dietary Fiber 5.2g   19%
Total Sugars 25.1g
Protein 5.4g

Stuffed Shells

Preparation Time: 15 Minutes
Cooking Time: 20 Minutes
Yield: 2 Servings

Ingredients
1 package dry pasta jumbo shells,
2 pounds ricotta cheese
1/2 bag spinach
1 jar marinara sauce
1/2 cup grated Parmesan cheese

Directions
Grease a baking sheet with oil spray
Stuff the shells with ricotta cheese.
Then arrange it on a sheet pan, cheese-side up.
In a shallow bowl mix spinach with marinara.
Put it on top of shells.
Select BAKE functions of CalmDo Air Fryer Oven, and adjust the time to 375 degrees F for 20 minutes
Place sheet pan inside a CalmDo Air Fryer Oven

Press START/PAUSE to begin.
When cooking is complete, remove and let it and top it with grated parmesan.
Serve.

Nutrition Facts
Servings: 2
Amount per serving
Calories   196
% Daily Value*
Total Fat 8.4g   11%
Saturated Fat 1.8g   9%
Cholesterol 45mg   15%
Sodium 1030mg 45%
Total Carbohydrate 11g   4%
Dietary Fiber 4.2g   15%
Total Sugars 2.3g
Protein 20g

Roasted Green Beans

Preparation Time: 10 Minutes
Cooking Time: 20-25 Minutes
Yield: 2 Servings

Ingredients
2 slices prosciutto
½ pound green beans, ends trimmed
½ small yellow onion, sliced
½ tablespoon canola oil

Directions
Take an air fryer basket and put prosciutto in it
Select air fry and set the timer to 4 minutes at 400 degrees

Let the preheating begin of CalmDo Air Fryer Oven.
Then out basket in air fryer and bake for 4 minutes.
In a bowl mix all the remaining ingredients
Afterward, take out prosciutto.
Put the vegetables in CalmDo Air Fryer Oven
Roast it for 15 minutes
Serve by crumbling the prosciutto on top of roasted green beans
Enjoy

Nutrition Facts

Servings: 2
Amount per serving
Calories   143
% Daily Value*
Total Fat 7.7g   10%
Saturated Fat 1.8g   9%
Cholesterol 25mg   8%
Sodium 648mg 28%
Total Carbohydrate 9.7g   4%
Dietary Fiber 4.2g   15%
Total Sugars 2.3g
Protein 11.2g

Spinach And Broccoli

Preparation Time: 10 Minutes
Cooking Time: 10 Minutes
Yield: 2 Servings

Ingredients

2 cups of spinach leaves, stem removed
1 cup broccoli florets
2 tablespoon of olive oil
Salt, to taste
Black pepper, to taste
1 teaspoon of onion powder

Directions
Rinse and pat dry the spinach leaves and remove the stems.
Take a large bowl and toss spinach leaves with broccoli, oil, salt, black pepper, and onion powder
Toss the ingredients well for fine coating
Oil greases the mesh basket and adds vegetables to it.
Set the timer of CalmDo Air Fryer Oven for 8-10 minutes at 390 degrees F.
Once the timer complete, take out the crisp takeout and serve. Enjoy.

Nutrition Facts
Servings: 2
Amount per serving
Calories   147
% Daily Value*
Total Fat 14.3g   18%
Saturated Fat 2g   10%
Cholesterol 0mg   0%
Sodium 117mg 5%
Total Carbohydrate 5.1g   2%
Dietary Fiber 1.9g   7%
Total Sugars 1.3g
Protein 2.3g

Blooming Onion

Preparation Time: 10 Minutes
Cooking Time: 22 Minutes
Yield: 2 Servings

Ingredients
1 large white onion
Salt and black pepper, to taste
2 eggs, whisked

4 tablespoons of olive oil
1 cup Panko bread crumbs
½ teaspoon of garlic powder
¼ teaspoon of paprika
1 cup flour
Oil spray, for coating

Directions
The first step is to peel the onions and cut off the top.
Afterward makes 4 slices of onion, till the bottom of the onion leaving about a centimeter.
Do not cut the way through.
The onion should be Bloom open.
Put the onions in the ice water for 4 hours
Take a bowl, and mix salt, garlic powder, paprika, black pepper, and flour.
Take out the onions from the ice water and pat dry.
Now dredge the onions into the flour mix.
Coat the onions well, and then pour the whisk. Egg over the onions.
In a small bowl combine breadcrumbs and olive oil.
Coat the onion with bread crumbs.
Take a mesh basket and coated with oil spray.
Add the onions in the mesh basket.
Put the mesh basket in CalmDo Air Fryer Oven
Set a timer to 22 minutes at 400 degrees Fahrenheit.
Once the onions are Crisp, take out and serve with any dipping sauce.

Nutrition Facts
Servings: 2
Amount per serving
Calories   779
% Daily Value*
Total Fat 36.2g   46%
Saturated Fat 6.2g   31%
Cholesterol 164mg   55%

Sodium 461mg 20%
Total Carbohydrate 94.6g 34%
Dietary Fiber 5.9g   21%
Total Sugars 7.2g
Protein 20.2g

French Fries

Preparation Time: 10 Minutes
Cooking Time: 22 Minutes
Yield: 2 Servings

Ingredients

500 grams white potatoes cut it into strips
Sea salt
2 tablespoons of olive oil
1 cup ketchup, side serving

Directions

Soak the potato strips in cold water and leave it for 1hours
Then pat dry the strips and let them sit for a while to get dry
Season it with salt and oil.
Put it in a rolling cage and use tongs to secure the position into the slot.
 Pace it in CalmDo Air Fryer Oven
Set timer to 22 minutes at 395degees F.
Once done, serve with ketchup.

Nutrition Facts

Servings: 2
Amount per serving
Calories   663
% Daily Value*
Total Fat 14.9g   19%
Saturated Fat 2.2g   11%
Cholesterol 0mg   0%

Sodium 1370mg 60%
Total Carbohydrate 126.9g   46%
Dietary Fiber 12.5g   45%
Total Sugars 31.6g
Protein 13.3g

Ginger Scones

Preparation Time: 10 Minutes
Cooking Time: 12-24 Minutes
Yield: 4 Servings

Ingredients
2.5 cups of all-purpose flour
1/2 cup of brown sugar
1 teaspoon of cinnamon
1 teaspoon of cloves
2 teaspoon of baking soda
12 tablespoons of butter
½ cup heavy cream
1 tablespoon of canal oil
1 egg, organic
1 tablespoon of vanilla extract
1 cup cranberries

Directions
Combine cranberries with all the dry ingredients
Whisk the egg in a bowl and add all the wet ingredients combine dry ingredients with wet ingredients.
Make the dough in a circle.
Cut the dough into 8 wedges
Brush the top with additional oil.
Put it in a CalmDo Air Fryer Oven rack in batches
Bake it in CalmDo Air Fryer Oven for 12 minutes at 180 degrees F

Once done, serve and enjoy

Nutrition Facts
Servings: 4
Amount per serving
Calories   685
% Daily Value*
Total Fat 42.1g   54%
Saturated Fat 25.8g   129%
Cholesterol 153mg   51%
Sodium 899mg   39%
Total Carbohydrate 63.9g   23%
Dietary Fiber 3.6g   13%
Total Sugars 1.8g
Protein 10.2g

Potato Tots

Preparation Time: 10 Minutes
Cooking Time: 10 Minutes
Yield: 4 Servings

Ingredients
10 potato tots
10 bacon strips
2 scallions
1 cup sour cream
Oil spray, for greasing

Directions
Wrap potato tots with a bacon strip and place them on a baking rack that is greased with oil spray.
Put the baking rack in CalmDo Air Fryer Oven.
Chose French fries and press start
Set timer to 10 minutes at 375 degrees F.

Once the cooking cycle complete, take out the potato tart and sprinkle sour cream and scallion on top.
 Serve and enjoy.

Nutrition Facts
Servings: 4
Amount per serving
Calories   752
% Daily Value*
Total Fat 51g   65%
Saturated Fat 12.5g   63%
Cholesterol 25mg   8%
Sodium 1457mg 63%
Total Carbohydrate 55.5g 20%
Dietary Fiber 7.7g   28%
Total Sugars 0.3g
Protein 12g

Chapter 6: Soups, Stews and Broth 9

Chicken Gravy

Preparation Time: 25 Minutes
Cooking Time: 20 Minutes
Yield: 3-4 Servings

Ingredients
2 tablespoons of olive oil
2 tablespoons soy sauce
1 teaspoon of ginger garlic paste
3 teaspoons poultry seasoning
¼ cup cream of mushroom soup
2 cups of chicken breasts, cubed
2 tablespoons of tomato paste
1/3cup coconut milk /cream

Directions
In a bowl mix soy sauce, olive oil, chicken breast pieces, tomato paste, poultry seasoning, and ginger garlic paste
Let it sit for 15mintus.
Preheat CalmDo Air Fryer Oven to 400 degrees F
Line the bottom of the CalmDo Air Fryer Oven baking pan with parchment paper.
Bake the chicken in it for 15 minutes at 400 degrees F.
Then add coconut milk and let it bake for additional 4 minutes
Serve over rice.

Steak And Mushroom Gravy

Preparation Time: 25 Minutes
Cooking Time: 10 Minutes
Yield: 3-4 Servings

Ingredients
1/3 cup olive oil
2 tablespoons coconut amino
3 teaspoons Montreal steak seasoning
1 teaspoon garlic powder
3 strip steaks cut into 3/4-inch pieces
¼ cup cream of mushroom soup

Directions
In a bowl mix coconut amino, olive oil, steak pieces, steak seasoning, and garlic powder
Let it sit for 15mintus
Preheat CalmDo Air Fryer Oven to 390 degrees F
Line the bottom of the air fryer baking pot with d parchment paper.
Put the steak with marinade.
Bake for 5 minutes
Then adds the cream of mushroom soup.
Put it in the air fryer and bake it for additional 5 minutes
Once done serve.
Nutrition Facts
Servings: 4
Amount per serving
Calories   279
% Daily Value*
Total Fat 23g   29%
Saturated Fat 4.6g   23%
Cholesterol 34mg   11%
Sodium 94mg   4%
Total Carbohydrate 1g   0%

Dietary Fiber 0.1g   0%
Total Sugars 0.3g
Protein 17.7g

Mongolian Beef

Preparation Time: 25 Minutes
Cooking Time: 8 Minutes
Yield: 3-4 Servings

Ingredients
1 pound of flank steak
1/4 cup corn starch

Sauce Ingredients
2 teaspoons of vegetable oil
1/2 teaspoon of ginger garlic paste
1/3 cup soy sauce or gluten-free soy sauce
1/3 cup water
1/4 cup brown sugar

Directions

Coat the steak with corn starch and put it in a baking tray and air fryer for 5 minutes at 400 degrees F, in CalmDo Air Fryer Oven.
In a skillet warm all the sauce ingredients.
Put the steak in the sauce and let it cook for 3 minutes.
Serve over rice.

Nutrition Facts
Servings: 4
Amount per serving
Calories   321
% Daily Value*
Total Fat 11.7g   15%

Saturated Fat 4.4g 22%
Cholesterol 62mg 21%
Sodium 1264mg 55%
Total Carbohydrate 19.5g 7%
Dietary Fiber 0.2g 1%
Total Sugars 9.2g
Protein 32.9g

Lamb Chops In Yogurt

Preparation Time: 25 Minutes
Cooking Time: 20 Minutes
Yield: 2 Servings

Ingredients
1 cup low-fat yogurt, side serving
½ teaspoon of cumin powder
½ tablespoon of coriander powder
1/3 teaspoon chili powder
½ teaspoon Gram Masala powder
1 tablespoon lemon juice
¼ teaspoon salt
Black pepper, to taste
4 lamb chops, bone-in

Directions
In a large bowl whisk yogurt with salt, cumin powder, pepper, coriander powder, chili powder, lemon juice, and grams Masala powder.
Marinate the chops in it for 30 minutes.
Afterward, place chops in a CalmDo Air Fryer Oven baking pan and bake for 20 minutes.
Remember to flip halfway through.

Once done serve

Nutrition Facts

Servings: 2

Amount per serving

Calories 1309

% Daily Value*

Total Fat 49.7g 64%

Saturated Fat 18.4g 92%

Cholesterol 596mg 199%

Sodium 880mg 38%

Total Carbohydrate 9.3g 3%

Dietary Fiber 0.3g 1%

Total Sugars 8.8g

Protein 190.8g

Coconut Salmon Gravy

Preparation Time: 10 Minutes
Cooking Time: 15 Minutes
Yield: 2 servings

Ingredients

1 pound salmon fillet
2 tablespoons olive oil
¼ teaspoon of paprika
1/3 teaspoon of cumin
½ teaspoon of red chili flakes
Salt and black pepper
½ cup of coconut milk
2tablespoonsoftaotme paste
1onion, chopped
Salt and black pepper, to taste

Directions

Take a baking tray or pan and cover it with Aluminum foil.

Rub the salmon fillet with the listed spices.
Put Salmon in the CalmDo Air Fryer Oven and bake for 10 minutes at 390
In a skillet add oil and cook the onion for 3 minutes
Then add salt, black pepper and tomatoes paste along with coconut milk
Once salmon is cooked.
Serve it with coconut tomato gravy.

Nutrition Facts
Servings: 2
Amount per serving
Calories   582
% Daily Value*
Total Fat 42.5g   54%
Saturated Fat 16.7g   84%
Cholesterol 100mg   33%
Sodium 112mg   5%
Total Carbohydrate 8.8g   3%
Dietary Fiber 2.7g   9%
Total Sugars 4.4g
Protein 46.1g

Tomato Soup

Preparation Time: 25 Minutes
Cooking Time: 4 Minutes
Yield: 4 Servings

Ingredients
1 cans tomatoes soup
¼ cup of coconut milk
2 basil leaves
Salt and black pepper

Directions

Take a heatproof pot and put the listed ingredient in it
Put it inside a CalmDo Air Fryer Oven and air fry for 4 minutes at 390 degrees F.
Serve hot.

Nutrition Facts

Servings: 4
Amount per serving
Calories   80
% Daily Value*
Total Fat 4g   5%
Saturated Fat 3.3g   16%
Cholesterol 0mg   0%
Sodium 422mg   18%
Total Carbohydrate 11.1g 4%
Dietary Fiber 1.3g   5%
Total Sugars 6.7g
Protein 1.6g

Vegetable Stew

Preparation Time: 25 Minutes
Cooking Time: 40 Minutes
Yield: 3 Servings

Ingredients

5 tablespoons olive oil
1 pound potatoes, halved and sliced
1 teaspoons salt, divided
1 zucchini, halved and sliced
1 leek, thinly sliced
4 stalks celery, sliced
10 ounces baby Bella mushrooms, quartered

4 cups frozen artichokes, thawed
1 (15 ounces) diced tomatoes, canned
1 cup Parmesan cheese rind
½ teaspoon ground pepper
4 cups of water

Directions

Preheat CalmDo Air Fryer Oven at 350 degrees F.
In a cooking pot pour the oil and add potatoes slices.
Season it with salt and add all the listed vegetables
Pour tomatoes over the vegetables
Add water and bring it to a boil with the lid on top.
Once boiling, transfer the pot to the CalmDo Air Fryer Oven.
Bake it for 3 0 at 310 degrees F until vegetables get tender
Add parmesan at the end and stir
Serve.

Nutrition Facts

Servings: 3
Amount per serving
Calories   395
% Daily Value*
Total Fat 24.7g   32%
Saturated Fat 3.8g   19%
Cholesterol 2mg   1%
Sodium 84mg   4%
Total Carbohydrate 41.4g 15%
Dietary Fiber 8.5g   30%
Total Sugars 9.8g
Protein 8.3g

Madagascan Bean Stew

Preparation Time: 25 Minutes
Cooking time: 40-45 Minutes
Yield: 3-4 Servings

Ingredients
10 ounces of baby new potatoes
2 teaspoon of olive oil
1 onion, finely diced
1 cup black beans, drained
2 cloves garlic, minced
1 teaspoon of ginger
2 tablespoon of tomatoes pure
Black pepper and salt, to taste
1 cup of vegetable stock
1/2 tbsp cornstarch+2 tablespoon of eater
½ cup of rocket (arugula)

Directions
Put the potato in a CalmDo Air Fryer Oven basket and air fry for 10 minutes at 220 degrees F.
Then add the onion, and cook for 5 minutes.
Next add garlic, beans, olive oil, ginger, tomato puree, and seasoning.
Pour in the vegetable stock and corn starch mixture.
Air dries it for 20 minutes.
Add the rocket and air fry for 5 more minutes.
Once done, serve over rice.

Nutrition Facts
Servings: 4
Amount per serving
Calories   202
% Daily Value*

Total Fat 3.1g   4%
Saturated Fat 0.5g   3%
Cholesterol 0mg   0%
Sodium 17mg   1%
Total Carbohydrate 34g   12%
Dietary Fiber 8.2g   29%
Total Sugars 2.5g
Protein 11.1g

Mushroom Stew

Preparation Time: 15 Minutes
Cooking Time: 25 Minutes
Yield: 3 Servings

Ingredients
3 tablespoons olive oil
1 cup tofu, cubed
salt, to taste
1 zucchini, halved and sliced
10 ounces baby Bella mushrooms, quartered
1 (15 ounces) diced tomatoes, canned
½ teaspoon red chili powder
4 cups of coconut milk

Directions
Preheat CalmDo Air Fryer Oven at 350 degrees F for 3minutes.
In a cooking pot add the olive oil and add mushrooms and tomatoes.
Season it with salt and add zucchini ,salt ,and red chili powder.
Add coconut milk and bring it to a boil with the lid on top.
transfer it to the cooking pot and add tofu
Put it in the CalmDo Air Fryer Oven.
Bake it for 20 minutes at 310 degrees F until vegetables get tender

then Serve over rice.

Nutrition Facts

Servings: 3

Amount per serving

Calories   977

% Daily Value*

Total Fat 94.5g   121%

Saturated Fat 70.5g   352%

Cholesterol 0mg   0%

Sodium 132mg   6%

Total Carbohydrate 31.9g   12%

Dietary Fiber 11.8g   42%

Total Sugars 17.8g

Protein 18.7g

Chapter 7: Beans and Eggs 9

Kidney Beans

Preparation Time: 25 Minutes
Soak Time: 31 Minutes
Yield: 3-4 Servings

Ingredients
1.5 cups kidney beans, canned and drained
1 large red onion, chopped
Oil spray, for greasing the pan
3 garlic cloves minced or paste
1 inch or teaspoon ginger, paste
Salt, to taste
½ teaspoon of turmeric
½ teaspoon of Red chili powder, to taste
4 cups of water

Directions
Soak the beans for two hours before starting the cooking.
Pour oil in a CalmDo Air Fryer Oven baking pan along with all the ingredients excluding beans.
Bake it for 6 minutes at 360 degrees F.
Then add beans and let it cook for 25 minutes.
Serve it whole wheat bread or rice.

Nutrition Facts
Servings: 4
Amount per serving
Calories   251
% Daily Value*
Total Fat 1g   1%

Saturated Fat 0.2g 1%
Cholesterol 0mg 0%
Sodium 59mg 3%
Total Carbohydrate 46.2g 17%
Dietary Fiber 11.5g 41%
Total Sugars 3.1g
Protein 16g

Garbanzo Beans

Preparation Time: 15 Minutes
Soak Time: 25 Minutes
Yield: 3 Servings

Ingredients
1.5 cups garbanzo beans, canned and drained
2 large red onion, chopped
2 large tomatoes
2 tablespoons of olive oil
2 garlic cloves , paste
1.5 inch or teaspoon ginger, paste
½ teaspoon of turmeric
½ fenugreek powder
½ teaspoon of Red chili powder, to taste
4 cups of water
salt, to taste
1 teaspoon of Garam Masala

Directions
Take a skillet and heat oil in it
Sauté onions in it then add tomatoes
Cook it for 10 minutes
Then transfer it to a bakign pan.

Add green chilies
Mix garbanzo beans with salt, red chili, fenugreek powder, ginger, garlic, and red chili powder.
Mix well and add water
Now pour this into the baking pan.
Stir the ingredients
Bake it for 15 minutes at 360 degrees F in CalmDo Air Fryer Oven. once done, take out and serve.

Nutrition Facts
Servings: 3
Amount per serving
Calories   511
% Daily Value*
Total Fat 15.8g   20%
Saturated Fat 2g   10%
Cholesterol 0mg   0%
Sodium 100mg 4%
Total Carbohydrate 75.9g 28%
Dietary Fiber 21.3g   76%
Total Sugars 18.2g
Protein 21.7g

Navy Beans In Air Fryer

Preparation Time: 25 Minutes
Soak Time: 32 Minutes
Yield: 3-4 Servings

Ingredients
1.5 cups navy beans, drained
1 large white onions, chopped
Oil, for greasing the pan
2 garlic cloves minced or paste
1 inch or teaspoon ginger, paste

Salt, to taste
1/3 teaspoon of turmeric
1 teaspoon of red chili powder, to taste
Black pepper, to taste
4 cups of water

Directions
Preheat the CalmDo Air Fryer Oven at 360 degrees For 3 minutes.
Pour oil in an air fryer baking pan along with onions, ginger, garlic, turmeric, chili, black pepper, and salt.
Cook it for 10 minutes.
Then open the CalmDo Air Fryer Oven.
Add the beans along with water.
Cook for 22 minutes.
The bean gets tender.
Serve and enjoy.

Nutrition Facts
Servings: 3
Amount per serving
Calories   374
% Daily Value*
Total Fat 1.8g   2%
Saturated Fat 0.2g   1%
Cholesterol 0mg   0%
Sodium 76mg   3%
Total Carbohydrate 68.5g 25%
Dietary Fiber 26.8g   96%
Total Sugars 6.2g
Protein 23.9g

Hard-Cooked Eggs

Preparation Time: 12 Minutes
Cooking Time: 12-15 Minutes
Yield: 3 Servings

Ingredients
4 large eggs, organic
Salt and black pepper, to taste
4cups of water

Directions
Put the 4 eggs in a CalmDo Air Fryer Oven basket along with 4 cups of water
Air fry at 360 degrees for 12-15 minutes
Transfer eggs to very cold water.
Peel and serve with a sprinkle of salt and pepper.

Nutrition Facts
Servings: 3
Amount per serving
Calories   95
% Daily Value*
Total Fat 6.6g   9%
Saturated Fat 2.1g   10%
Cholesterol 248mg   83%
Sodium 103mg 4%
Total Carbohydrate 0.5g   0%
Dietary Fiber 0g 0%
Total Sugars 0.5g
Protein 8.4g

Eggs In A Hole

Preparation Time: 12 Minutes
Cooking Time: 10 Minutes
Yield: 2 Servings

Ingredients
2 slices of bread
2 tablespoon of butter
Salt and pepper to taste
2 eggs, organic

Directions
Make holes in the middle of bread slices with glass.
Butter the CalmDo Air Fryer Oven pan.
Put the bread slice into CalmDo Air Fryer Oven pan and crack one egg into a hole. Sprinkle salt and pepper.
Bake to 360 degrees F for 6 minutes.
Remove the bread slice and flip the bread and place it into the CalmDo Air Fryer Oven again.
Cook for 5 more minutes. Serve.

Nutrition Facts
Servings: 2
Amount per serving
Calories   189
% Daily Value*
Total Fat 16.2g   21%
Saturated Fat 8.7g   44%
Cholesterol 194mg   65%
Sodium 205mg   9%
Total Carbohydrate 4.9g   2%
Dietary Fiber 0.2g   1%
Total Sugars 0.7g
Protein 6.4g

Easy Breakfast Sandwich

Preparation Time: 12 Minutes
Cooking Time: 8 Minutes
Yield: 2 Servings

Ingredients
2 organic eggs
2 English bacons
2 streaky bacon slices
2 English muffin
Black pepper and salt, to taste (for sprinkling)

Directions
Take a soufflé cups and crack an egg in it.
 Put the egg cup, bacon, and English muffin into the CalmDo Air Fryer Oven and cook for 400 degrees F for 8 minutes.
Assemble all the ingredients in the form of a sandwich
Serve hot.

Nutrition Facts
Servings: 2
Amount per serving
Calories   598
% Daily Value*
Total Fat 25.7g   33%
Saturated Fat 9.4g   47%
Cholesterol 415mg   138%
Sodium 1504mg 65%
Total Carbohydrate 55.2g 20%
Dietary Fiber 2.6g   9%
Total Sugars 4.9g
Protein 36.2g

Air Fryer Crispy Chickpeas

Preparation Time: 12 Minutes
Cooking Time: 14 Minutes
Yield: 3 Servings

Ingredients
1 15 ounces can garbanzo beans, drained
1 teaspoon chili powder
1 teaspoon cumin, ground
Salt, to taste
1/4 teaspoon dry mustard
¼ teaspoon Smoked Paprika
¼ teaspoon garlic powder
1 teaspoon of Hidden Valley Dry Seasoning

Directions
Preheat CalmDo Air Fryer Oven to 390 degrees Fahrenheit for 20 minutes.
Rinse and drain garbanzo beans.
Transfer it to a paper towel let it dry
Then put it in the CalmDo Air Fryer Oven.
Air fries it for 2 minutes in calmdo air fryer at 390 degrees.
Open the basket and spray the chickpeas with oil spray
Air Fry for 10 more minutes.
Then open the calmdo air fryer and all seasoning the basket
Shake the basket and then air fry for 2 more minutes
Once chickpeas get crispy, serve
Serve immediately. Store any leftovers in a brown paper bag.

Nutrition Facts
Servings: 3
Amount per serving
Calories   524
% Daily Value*
Total Fat 9g   12%

Saturated Fat 0.9g   5%
Cholesterol 0mg   0%
Sodium 95mg   4%
Total Carbohydrate 87.1g 32%
Dietary Fiber 25.2g   90%
Total Sugars 15.4g
Protein 27.7g

Omelet In Air Fryer

Preparation Time: 12 Minutes
Cooking Time: 15 Minutes
Yield: 2 Servings

Ingredients

4 eggs, organic
4 tablespoons of butter
1/4 cup coconut milk
1/3 cup cheese, grated
Salt and black pepper, to taste
1 green onion, chopped

Directions

Take a shallow bowl and beat the eggs in it.
Add the butter and coconut milk.
Add grated cheese and mix well
Season it with salt, pepper, and add chopped green onion
Whisk together all the ingredients.
Preheat the CalmDo Air Fryer Oven to 200 degrees for 5 minutes.
Pour the omelet mixture into ramekins.
Set the CalmDo Air Fryer Oven to 180 degrees and cook the mixture in ramekins for 12-18 minutes.
Once egg gets firm, serve.

Nutrition Facts

Servings: 2

Amount per serving
Calories   477
% Daily Value*
Total Fat 45.2g   58%
Saturated Fat 27.6g   138%
Cholesterol 408mg   136%
Sodium 409mg 18%
Total Carbohydrate 3.2g   1%
Dietary Fiber 0.9g   3%
Total Sugars 2g
Protein 16.8g

Mushroom Omelet In Air Fryer

Preparation Time: 12 Minutes
Cooking Time: 15 Minutes
Yield: 2 Servings

Ingredients
2 eggs, organic
1 tablespoon of butter, melted
1/4 cup coconut milk
1/4 cup cheese, grated
4 mushrooms, chopped
2 tablespoons of bell pepper
Salt and black pepper, to taste

Directions
Take a bowl and beat the eggs in it.
Add the bell pepper, coconut milk, cheese, mushrooms, salt, black pepper, and melted butter
Whisk together all the ingredients.
Preheat the CalmDo Air Fryer Oven to 200 degrees for 5 minutes.
Place the omelet mixture into ramekins.

Set the CalmDo Air Fryer Oven to 180 degrees and cook the mixture in ramekins for 15 minutes.

Once eggs get firm, serve.

Nutrition Facts

Servings: 2

Amount per serving

Calories   248

% Daily Value*

Total Fat 22.1g   28%

Saturated Fat 14.3g   72%

Cholesterol 194mg   65%

Sodium 197mg 9%

Total Carbohydrate 3.4g   1%

Dietary Fiber 1g 4%

Total Sugars 2g

Protein 11g

Chapter 8: Desserts And Snacks 9

Zucchini Crisps

Preparation Time: 10 Minutes
Cooking Time: 18 Minutes
Yields: 2 Servings

Ingredients
8-10 tablespoons of water
Sea Salt, to taste
Paprika, pinch
Red chili flakes, to taste
1-2 zucchinis, peeled, round sliced
½ cup chickpea flour
Oil spray, for greasing

Directions
In a large bowl combine sea salt paprika, red chili flakes, water, and chickpea flour.
Add the sliced zucchini into the chickpea flour mixture to coat it well.
Grease the CalmDo Air Fryer Oven basket with oil spray.
Add the zucchini to the air fryer basket.
Cook it for about 15-18 minutes at 390 degrees F.
During cooking, shake it well.
Remember not to overlap it.
Serve with your favorite sauce.

Nutrition Facts
Servings: 2
Amount per serving
Calories   200
% Daily Value*

Total Fat 3.5g   4%
Saturated Fat 0.4g   2%
Cholesterol 0mg   0%
Sodium 141mg 6%
Total Carbohydrate 33.6g 12%
Dietary Fiber 9.8g   35%
Total Sugars 7.1g
Protein 10.8g

Apple Crisp

Preparation Time: 10 Minutes
Cooking Time: 60 Minutes
Yield: 2 Servings

Ingredients
2 apples, peeled and center cored
1 tablespoon of brown sugar
1/3 teaspoon cinnamon
Pinch of salt
Oil spray for greasing

Directions
Peel and center core the apples.
Use a mandolin to cut the apples into thin slices.
Take a bowl and mix apples, cinnamon, salt, and brown sugar.
Grease the CalmDo Air Fryer Oven mesh basket with oil spray.
Put the slices in the mesh basket.
Set a timer to 60 minutes at 283 degrees F.
Once crisp, takeout the apple crisps and let it get cool before serving.

Nutrition Facts
Servings: 2
Amount per serving
Calories   137

% Daily Value*
Total Fat 0.7g 1%
Saturated Fat 0g 0%
Cholesterol 0mg 0%
Sodium 81mg 4%
Total Carbohydrate 35.5g 13%
Dietary Fiber 5.6g 20%
Total Sugars 27.6g
Protein 0.6g

Walnut Brownies

Preparation Time: 10 Minutes
Cooking Time: 22 Minutes
Yield: 2 Servings

Ingredients
60 grams chocolate
80 gram butter
1 small egg
50 grams of brown sugar
1 teaspoon of vanilla essence
55 grams of flour
30 gtams walnuts, chopped

Directions
The first step is to combine butter and chocolate in a small bowl and heat in a microwave for 2 minutes.
In a mixing bowl, whisk eggs, vanilla essence, and brown sugar.
Add chocolate to the egg mixture, and fold in the flour and walnuts.
Line a Cake tin with baking paper and fill it with the mixture.
Set the temperature of the CalmDo Air Fryer Oven to 20 minutes at 350 degrees Fahrenheit.

Put the small cake tin inside the CalmDo Air Fryer Oven, and bake it until the top is crispy and brown.
Take out the cake tin and let it cool.
Then serve.

Nutrition Facts
Servings: 2
Amount per serving
Calories   868
% Daily Value*
Total Fat 61.9g   79%
Saturated Fat 28.4g   142%
Cholesterol 162mg   54%
Sodium 288mg   13%
Total Carbohydrate 66.9g 24%
Dietary Fiber 3.9g   14%
Total Sugars 40.6g
Protein 15.4g

Kale Chips

Preparation Time: 10 Minutes
Cooking Time: 5 Minutes
Yield: 2 Servings

Ingredients

2 cups of kale
1 tablespoon of olive oil
Salt, to taste

Directions

Rinse and pat dry the kale.
Remove the stem of the kale.

In a bowl mix kale, olive oil, and salt
Toss the ingredients well.
Oil greases the CalmDo Air Fryer Oven mesh basket and adds kale to it
Set the timer to 5 minutes at 400degrees F.
Once chip gets crisp takeout and serve.
Enjoy.

Nutrition Facts
Servings: 2
Amount per serving
Calories   93
% Daily Value*
Total Fat 7g   9%
Saturated Fat 1g   5%
Cholesterol 0mg   0%
Sodium 107mg 5%
Total Carbohydrate 7g   3%
Dietary Fiber 1g 4%
Total Sugars 0g
Protein 2g

Simply Sweet Desert

Preparation Time: 25 Minutes
Cooking Time: 10 Minutes
Yield: 3-4 Servings

Ingredients
1-1/3 cups all-purpose flour
1 teaspoon of baking powder
Pinch of salt
3 tablespoons white sugar
1/3 cup buttermilk
1 egg
5 tablespoons butter, melted

Icing sugar, garnishing

Directions

In a bowl, mix baking powder, salt, sugar, and flour
Mix it all well.
Then whisk the egg in a separate bowl and add buttermilk to the egg
Whisk it well
Add egg mixture to the flour mixture and combine all the ingredients.
Pat the dough on to clean work surface and roll flat.
Cut it into shapes and then brush melted butter on top
Put it to the CalmDo Air Fryer Oven baking tray and bake for 9 minutes at 392 degrees F.
Once puffy, take out and sprinkle icing sugar on top.
Serve and enjoy.

Nutrition Facts

Servings: 4
Amount per serving
Calories   300
% Daily Value*
Total Fat 16g   21%
Saturated Fat 9.6g   48%
Cholesterol 80mg   27%
Sodium 180mg 8%
Total Carbohydrate 34.5g 13%
Dietary Fiber 0.9g   3%
Total Sugars 10.2g
Protein 5.4g

Cheddar Biscuits

Preparation Time: 25 Minutes
Cooking Time: 15 Minutes
Yield: 3 Servings

Ingredients
1 cup flour
1 stick of butter
½ cup of scallion
1.5 cup cheddar cheese
1 cup buttermilk
2 teaspoon of baking soda
1 teaspoon of seafood seasoning

Direction
Mix flour with butter in a bowl.
Then add the remaining listed ingredients one by one
Mix and incorporate all the ingredients.
Divide the mixture into 8 equal balls and place it on one baking sheet.
Put it in CalmDo Air Fryer Oven, and select bake function.
Set time to 15minutes at 385 degrees. Serve and enjoy.

Nutrition Facts
Servings: 3
Amount per serving
Calories   687
% Daily Value*
Total Fat 50.4g   65%
Saturated Fat 31.8g   159%
Cholesterol 144mg   48%
Sodium 1497mg 65%
Total Carbohydrate 37.7g 14%
Dietary Fiber 1.6g   6%
Total Sugars 4.7g
Protein 21.7g

Banana And Chocolate Cups

Preparation Time: 30 Minutes
Cooking Time: 22-44 Minutes
Yield: 4 Servings

Ingredients
1.5 cups rolled oats or old fashioned oats
1/4 teaspoon ground cinnamon
Pinch ground nutmeg
½ teaspoon baking powder
Pinch of salt
1 large egg
1/8 cup monk fruit sweetener
1/2 cup mashed bananas
1 teaspoon pure vanilla extract
½ cup milk
1/8 cup melted coconut oil
½ cup mini chocolate chips

Directions
Preheat the CalmDo Air Fryer Oven to 350 degrees F.
Take a shallow bowl and combine nutmeg, cinnamon baking powder, salt, and rolled oats.
In a separate bowl, whisk the egg and add sweetener, vanilla extract, and bananas.
Then pour in milk and coconut oil into the egg mixture.
Now combine ingredients of both the bowl.
Dump the chocolate chips at the end.
Mix well and put the generous amount into grease muffin pan that adjusts in CalmDo Air Fryer Oven.
You can do the baking in batches.
Bake muffins in the calm do air fryer basket, for 22 minutes.
Once brown, let it cool for 15 minutes and serve.

Nutrition Facts

Servings: 4
Amount per serving
Calories   507
% Daily Value*
Total Fat 37.1g   48%
Saturated Fat 28.8g   144%
Cholesterol 52mg   17%
Sodium 79mg   3%
Total Carbohydrate 38.7g 14%
Dietary Fiber 4.4g   16%
Total Sugars 13.7g
Protein 7.5g

Mixed Nuts

Preparation Time: 10 Minutes
Cooking Time: 15 Minutes
Yield: 4 Servings

Ingredients
½ cup cashew
1 cup almond
1 cup peanut
¼ cup honey
Pinch of sea salt
¼ teaspoon of cinnamon
½ cup brown sugar

Directions
Take a shallow bowl and mix ingredients in it.
Transfer the nuts to the roasting basket of CalmDo Air Fryer Oven/
Select steak/chop functions of CalmDo Air Fryer Oven
Start cooking by adjusting the time to 15 minutes.
Once done, take it out and let it cool.

Then serve and enjoy.

Nutrition Facts

Servings: 4
Amount per serving
Calories   576
% Daily Value*
Total Fat 37.8g   48%
Saturated Fat 5g   25%
Cholesterol 0mg   0%
Sodium 74mg   3%
Total Carbohydrate 51.9g   19%
Dietary Fiber 6.7g   24%
Total Sugars 38.3g
Protein 17.2g

Bake Broccoli

Preparation Time: 10 Minutes
Cooking Time: 12 Minutes
Yield: 2 servings

Ingredients

2 cups broccoli florets
Salt and black pepper, to taste
1 tablespoon of olive oil
1 cup parmesan, shaved

Directions

Rinse and cut the broccoli florets.
Drizzle the olive oil on top.
And season it with salt and black pepper.
Oil sprays the mesh basket of CalmDo Air Fryer Oven.
Add florets to the mesh basket.
Set a timer to 12 minutes at 400 degrees F.

Serve with a sprinkle of parmesan shaves.

Nutrition Facts

Servings: 2

Amount per serving

Calories 361

% Daily Value*

Total Fat 25.3g 32%

Saturated Fat 13g 65%

Cholesterol 60mg 20%

Sodium 810mg 35%

Total Carbohydrate 9.1g 3%

Dietary Fiber 2.4g 9%

Total Sugars 1.6g

Protein 29.6g

Conclusion

The Calmdo air fryer oven is a new technology for making some crispy food. It is no doubt a remarkable appliance that can be operated easily with its LCD and one-touch technology. The functions are easy to use with easy time and temperature adjustments.

Now you can prepare some best meals in your home without going or ordering from restaurants.
It's efficient circulating and heating technology gives some astonishing result and aromatic meals along with mouthwatering flavors

To all cooking needs, users can now select any of the 18 smart programs. If you love cooking and preparing food then this appliance is best to buy as its flexibility of adjusting temperatures and settings is best.

Manufactured by Amazon.ca
Bolton, ON